Historic England

Civil Defence: From the First World War to the Cold War

Introductions to Heritage Assets

Summary

Historic England's Introductions to Heritage Assets (IHAs) are accessible, authoritative, illustrated summaries of what we know about specific types of archaeological site, building, landscape or marine asset. Typically they deal with subjects which lack such a summary. This can either be where the literature is dauntingly voluminous, or alternatively where little has been written. Most often it is the latter, and many IHAs bring understanding of site or building types which are neglected or little understood. Many of these are what might be thought of as 'new heritage', that is they date from after the Second World War.

Until the 20th century, neither the capacity nor the means to inflict major damage upon the British Isles existed. This situation changed during the First World War, with naval bombardments and aerial bombing of towns and cities by Zeppelin airships and Gotha bombers. These raids demonstrated that no one was immune from attack and that by developing civil defence measures, casualties could be reduced and lives saved. The range of civil defence buildings used in the Second World War was immense, most were built 'for the duration', often utilitarian both in design and materials, but none the less, historically and technologically significant.

This guidance note has been written by Roger J. C. Thomas and edited by Paul Stamper.

It is one is of several guidance documents that can be accessed at
HistoricEngland.org.uk/listing/selection-criteria/listing-selection/ihas-buildings/

Published by Historic England October 2016.

HistoricEngland.org.uk/listing/selection-criteria/

Front cover
A poster produced in 1942 extolling the virtues and benefits of using surface air raid shelters.

Contents

Introduction

Civil Defence (CD) or Air Raid Precautions (ARP) can be summed up as the provision of measures to protect the civil population from enemy air, naval, or artillery assault, whether by high-explosive, incendiary, chemical, biological, or nuclear weapons. These measures include command and control, warning, risk reduction, physical protection in either the home, work place or in public places, and evacuation. They also include fire and rescue, cleansing, medical care, emergency feeding, temporary accommodation, emergency repairs, clearance of debris, repair of public utilities, law enforcement, recovery of valuable materials and possessions, and the care of animals. Most surviving examples of structures associated with civil defence date predominantly from the Second World War and Cold War.

Most purpose-built structures have a utilitarian appearance; nevertheless, they embody historic and technical values illustrating the all-encompassing nature of total warfare and the experience of the civilian in conflict, and can be seen by local communities as 'touch-stones' with their past, and as an educational assets. With adapted structures, it is often very difficult without documentary evidence to identify alterations which were generally later removed; however, blast walls, partially bricked-up window spaces, reinforced concrete cellar steps canopies, cellar escape hatches, signage, strengthened floors, walls, and basements can still be discovered, as for example at Historic England's office at Tanner Row, York. These alterations are highly vulnerable to removal and are frequently overlooked due to a lack of recognition or appreciation; nevertheless, such modifications may contribute to the overall historic fabric and interest of a structure.

The study of civil defence structures is in its infancy, but there is a growing interest and appreciation that these buildings play an important role in our modern national heritage. They represent a physical manifestation of a civilian service that suffered 2,739 killed and 4,459 seriously injured while in the line of duty.

In the descriptions which follow, imperial measurements are largely used, following original specifications.

1 Historical Background

As an island nation, the British have generally felt safe from foreign attack. This illusion persisted until on the morning of 16 December 1914 when the battlecruisers of the German High Seas Fleet bombarded Scarborough, Whitby and Hartlepool, resulting in 137 fatalities and 592 other casualties. Six days later, what with hindsight was a highly significant event occurred at Dover, when a German seaplane dropped two bombs – as it happened harmlessly into the sea - marking the first attempted air attack on Great Britain. Two days later, the first bomb to land on British soil buried itself in Mr Tommy Terson's Dover garden and detonated amongst his vegetables. These attacks were the harbingers of much worse to come. Indeed the regular bombing of Paris was a clear indication of the German General Staff's intent to use air power to cause mass terror and the collapse of civilian morale.

Figure 1
Orderly queuing to enter an air raid shelter at Hither Green, Lewisham.

1.1 The First World War, 1914-19

The scale of attack escalated once the Germans embarked upon its Zeppelin bombing campaign against London on the night of 31 May / 1 June 1915. Civilian casualties and damage to property mounted sharply, causing an out cry over the lack of defences and the provision of shelters. By early 1917 it was clear that the Zeppelin campaign was faltering; however, a new and more challenging form of attack was about to commence. The first Gotha aeroplane bomber raid against London took place on 13 June 1917, when 14 aircraft reached the capital, dropping 118 bombs, killing 162 people, injuring 426, and causing £125,953 worth of damage. This scale of attack could no longer be ignored: the existing piecemeal provision of shelters in basements was reorganised on a district basis, the first purpose-built public air raid shelters were constructed, and an effective air raid warning system was introduced. Basements, cellars and crypts were strengthened as communal shelters, and in some places annexes were built onto the front of basement entrances. A single purpose-built First World War air raid shelter has been identified at Cleethorpes (Lincolnshire). Others are known through documentary and photographic sources (Fig 1).

1.2 Inter-war, 1919-39

At the end of the war civil defence was virtually abandoned. However, in 1924 the Committee for Imperial Defence established two Air Raid Precautions (ARP) sub-committees, one to consider policy and the other organisation. While war seemed unlikely civil discontent was seen as a threat, and a system emergency government was established that would later form the basis of the civil defence network. In terms of enemy attack from the air, many shared the view of Stanley

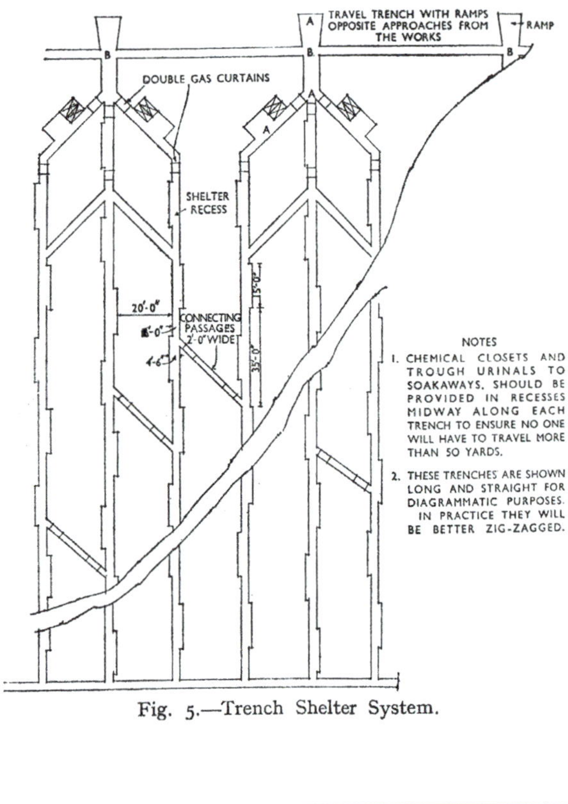

Figure 2
Air Raid Precautions Handbook No.6, suggested plan for a large-scale industrial trench shelter system, similar to that built by Harris Lebus, Tottenham Hale.

Figure 3
Interior of the concrete lined parabolic-profile trench shelter system built by Harris Lebus, Tottenham Hale, to protect some 2,500 workers.

Baldwin, then a member of Cabinet, who stood up in Parliament on 10 November 1932 saying:

> "I think it is well also for the man in the street to realise that there is no power on earth that can protect him from being bombed, whatever people may tell him. The bomber will always get through .... The only defence is offence, which means you have to kill more women and children more quickly than the enemy if you want to save yourselves."

Subsequently an appreciation of the need for air raid precautions grew, partially driven by the renouncement of the terms of the Treaty of Versailles by Nazi Germany, together with its aggressive policies and blatant re-armament By March 1935 a decision had been taken that local councils were best placed to provide ARP services, including first-aid, anti-gas, and a rescue service. In the following year, pressure was also brought to bear on large industrial concerns to provide similar protection to their workforce; needless to say, little was achieved without statutory compulsion. Concern over the protection of important industries resulted in the issuing of an official handbook in November 1936, *Air Raid Precautions in Factories and Business Premises*, which for the first time provided type-designs for structural precautions (Fig 2). While some large industrial companies such as Boots, Imperial Chemicals Industries and Harris Lebus of Tottenham Hale acted on the advice and built substantial shelters for their workforces the vast majority of business owners were reluctant to do so, chiefly due to cost.

The passing of the Air Raid Precautions Act in January 1938 resulted in local authorities being placed under a legal obligation to prepare ARP schemes. These emphasised 'dispersal' and 'self-reliance' rather than 'total' protection, and with little in the way of built fabric. The bombing of towns and cities during the Spanish Civil War, particularly Guernica, attracted the attention of politicians and scientists who favoured civil defence preparations. In 1938 Professor J.B.S. Haldane argued in his book *A.R.P.* (1938), contrary to the Government's line that it was technically feasible and cost effective to provide 'bomb-proof' shelters for the whole population. Never being a man to mince his words, the opening sentence reads: 'This book is intended for the ordinary citizen, the sort of man and woman who is going to be killed if Britain is raided again from the air.'

Initially, few took the ARP service seriously, and the only public guidance was a leaflet, *The Protection of your Home against Air Raids*. However, work was progressing with plans for 'blackout', air raid warning sirens, the distribution of gas masks, and proposals for evacuations, and the establishment of two Civil Anti-Gas Schools – Eastwood Park, Falfield (Gloucestershire), and Hawkhills, Easingwold (North Yorkshire). The Munich Crisis in the summer of 1938 suddenly brought the situation into focus: gas masks were distributed and shelter trenches dug in parks. However, once the immediate crisis had passed, the Government's lack of clear guidance on the design of shelters resulted in a number of commercial companies like Stent, Nissen, Concrete Limited (Bison), British Reinforced Concrete (BRC) and Allens of Tipton offering their own designs for private purchase.

In the meantime, the Civil Defence Emergency Scheme 'Y' was enacted in September 1938, with twelve Civil Defence regions being established, each with a Regional Commissioner. Approval was given in November 1938 for the improvised park trench shelters to be made permanent using pre-cast concrete trench linings, and basements and other buildings selected as public shelters to be strengthened. Approval was given in December for the issue of the first Government-designed domestic air raid shelter – the Anderson Shelter (Fig 4) – and for the supply and fitting (by the local authority) of steel strutting kits to strengthen basements as shelters. 2,300,878 Anderson shelters had been issued by September 1940, many free of charge to poorer families, and from May 1939, where conditions did not permit the use of an Anderson shelter, local authorities were empowered to build small brick and concrete surface shelters in backyards. Today these structures are often mistaken for outside

Figure 4 (top left)
An Anderson Shelter complete with earth revetment.

Figure 5 (top right)
A typical small domestic surface shelter built by York City Council in the backyard of a terraced house in the South Bank area of York, North Yorkshire.

Figure 6 (bottom)
Sayer Croft National Camp, Surrey.

lavatories or sheds, and examples can still be found associated with urban terraced housing, like those in South Bank, York (Fig. 5).

Although most building work during the twelve months after the Munich Crisis was of shelters, other projects included the erection of National Camps (Fig 6) to receive children evacuated from major cities, and some ARP Warden Posts. The Civil Defence Act 1939 finally came fully into force in July 1939, obliging local authorities to install domestic shelters, imposing ARP design considerations on new buildings, and finally placing a statutory responsibility on employers to provide ARP protection in the workplace.

1.3 The Second World War, 1939-45

On the outbreak of Second World War in September 1939, the Ministry for Home Security was created with Sir John Anderson as Home Secretary and Minister of Home Security. The Ministry was responsible for all central and regional civil defence organisations, including rescue, air raid wardens, fire, and the Women's Voluntary Service. It was also responsible for giving approval to local ARP schemes, and providing public shelters.

Initially, most of the civil defence services were housed in pre-existing structures. Subsequently, specific designs for all types of civil defence buildings were issued by the Ministry; however, considerable variety in detail and materials occurred as the local authorities adapted the official designs or drew up their own. These structures included domestic, public, communal, school, industrial and deep shelters; blast walls; ARP warden posts; control centres; war rooms; first aid posts; dressing stations; cleansing stations; gas decontamination centres (Fig 7); gas houses or chambers; ARP laundries; mortuaries; civil defence depots; ambulance garages; fire watchers' posts (Fire Guards); 'Jim Crow' posts (industrial bomb watchers); National Fire Service garages and fire stations; static water tanks (EWS emergency water supply); air raid warning siren posts and pillars; rest centres; national camps; buffer depots (dry food, frozen food, granaries, emergency supplies); and ARP gas schools and rescue training sets. Probably the most common type of structure built was the 50-person public shelter; it is thought that some 16,747 were built in the Leeds area alone. Alongside the new or adapted civil defence structures, painted, cast-iron and enamel signs pointing to air raid shelters, emergency water supplies and so on were installed. In addition, white patches, or black and

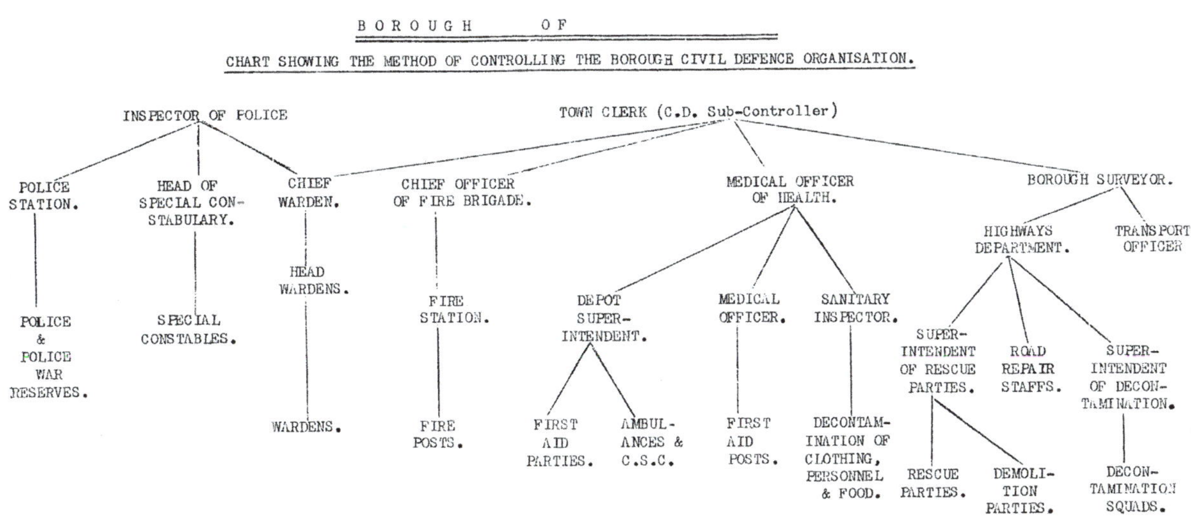

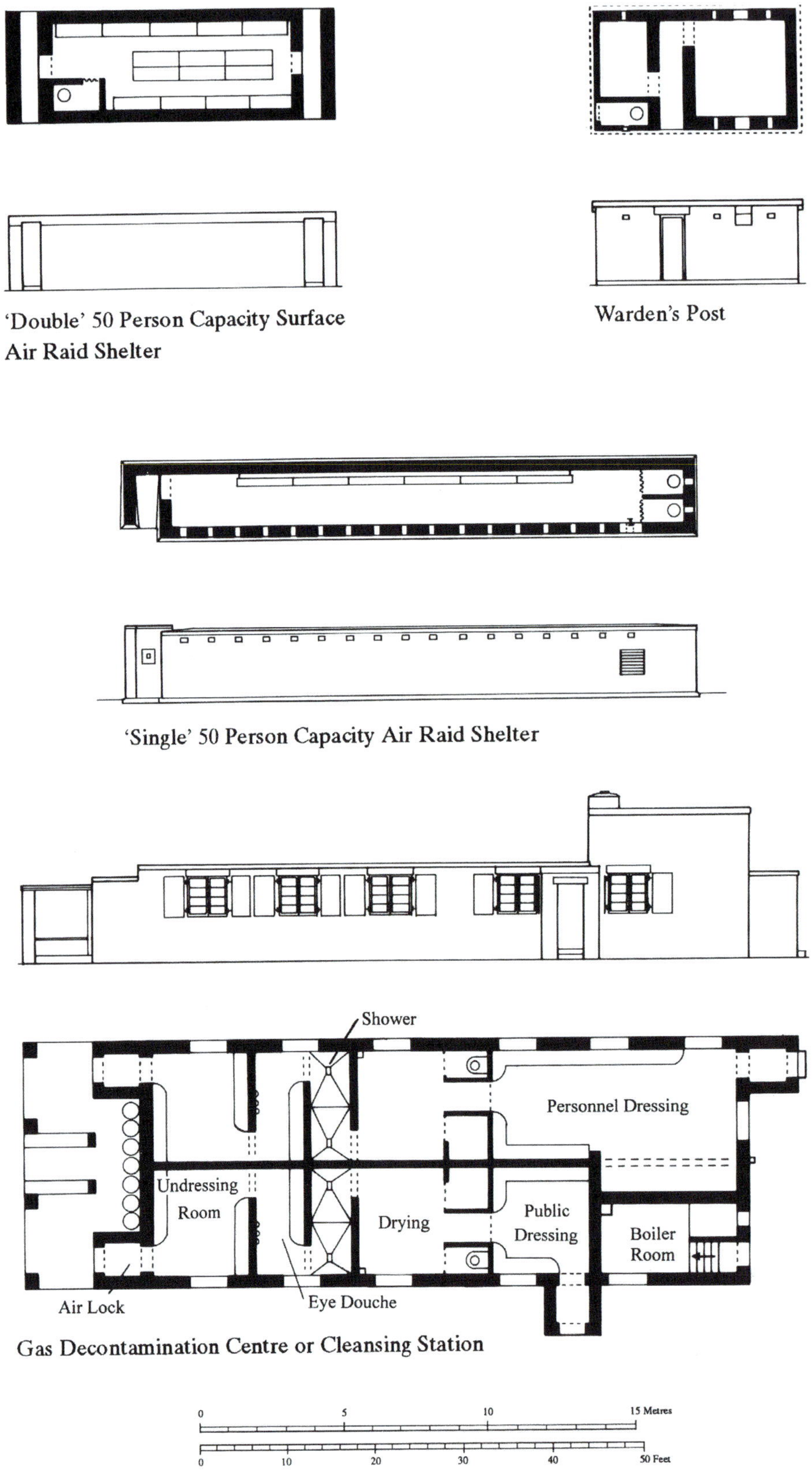

Figure 7
Examples of purpose-built civil defence structures.

Figure 8
A National Fire Service painted sign in St Cuthbert's Lane, Carlisle, pointing to the position of Emergency Water Supply No 5.

white striped markings on steps, pavements, lamp posts and doorways were painted to assist drivers and pedestrians in the 'blackout'; faded survivals are increasingly rare (Fig 8).

Air raid shelters ranged in size and complexity from the small Anderson shelter for four or six persons to substantial 'deep' public shelter complexes and underground industrial shelters designed for some 2,000 people. Shelters were also built at schools, and for families living near to military installations like airfields, barracks, depots, anti-aircraft batteries, or bombing decoys. Some such structures were later adapted for other uses, their original function perhaps forgotten. For example, at the Selby Abbey Church of England (VC) Primary School (North Yorkshire) a former surface air raid shelter is now used as a games equipment store, while the Westbourne Library, Ipswich (Suffolk; listed Grade II) occupies a former gas decontamination centre.

Although the authorities intended public shelters should only be occupied during a raid, and had expressly indicated that the London underground was not to be used for sheltering, the switch from day to night bombing by the Luftwaffe in October 1940 brought about the *de facto* occupation of many of London Underground tube stations as deep shelters, and the nightly 'Trekking' of urban populations into the countryside. Nightly migrations were feared to be having a detrimental effect on both morale and war industries, so a decision was taken to provide 'shelter colonies' on the edges of towns where trekking was taking place. However, by the time colonies were built, night-time bombing was tailing off.

While the main German blitz on British cities ended in May 1941, civil defence services were maintained as sporadic attacks and tip-and-run nuisance raids continued. It was during this period that many of the adapted pre-war buildings were replaced by purpose-built structures, and eight deep tube shelters were completed in London. The feared resumption of bombing raids came to pass in January 1944 – the so-called 'baby blitz' – and lasted until May 1944. By then, conscription and the needs of war industries had substantially reduced civil defence manpower. To overcome the problem, the Civil Defence Reserve was established, operating 'Mobile Columns' of rescue vehicles largely operating from council depots.

In the summer of 1944 Hitler's 'vengeance weapons' began to target London and the south-east: the pilotless V1 flying bomb or doodlebug (from 13 June), and the V2 long range ballistic rocket (from 8 September). This brought about a clamour for more Anderson and Morrison shelters, which was answered by re-claiming shelters from less threatened areas like South Wales. These attacks finally came to an end on 29 March 1945 as the advancing Allied armies over-ran the rocket launch sites.

The stand-down of the civil defence organisation followed soon afterwards, on 2 May 1945. The demolition of the vast majority of warden posts and the 50-person public shelters commenced virtually as soon as the local authorities obtained permission. Many stood on public roads, pavements and back lanes, forming hazards to vehicles and pedestrians. Where such structures stood on public authority land, they were often retained as storerooms for parks and schools, changing rooms at sports grounds (Fig 9), and some were even converted into public conveniences.

Figure 9
Westbourne Library, Ipswich, Suffolk is a former gas decontamination centre that was built in 1942 and was converted into a library between 1946 and 1947. Listed Grade II.

Figure 10
Epsom Civil Defence Training Set 4th May 1953.

The Government also attempted to recover the steel used in Anderson and Morrison shelters in basement strutting, but the lack of labour to carry out the recovery work resulted in householders being able to purchase their Anderson shelter for £1. The small brick and concrete shelters built as alternatives to Anderson shelters had no recovery value, and were usually given to the property owner.

1.4 The Cold War, 1948-89

At the stand-down in 1945 the Home Office had the foresight to encourage local authorities to fund the establishment of Civil Defence Associations, to maintain contact with former members. This was timely, given the growing tensions between the western allies and the Soviets. The Civil Defence Act 1948 established the Civil Defence Corps, a volunteer civilian organisation intended to manage and provide rescue services in areas affected by a major national emergency: after August 1949, when the Soviet Union conducted its first nuclear test that meant a nuclear attack.

Once again, Air Raid Wardens were recruited and the Civil Defence Corp adopted the 'Mobile Columns' of reconnaissance, communications and rescue vehicles. Civil defence depots were re-activated and some new depots were built. These were often associated with training grounds established in bombed out and shattered buildings, as for example at Epsom, Surrey (Fig 10). After the detonation of the Soviet hydrogen bomb in 1953, Government survival became the main priority. For a time during the 1950s those designing public building schemes were encouraged to incorporate shelters, but this had generally ended by 1960. As the perceived Soviet nuclear threat grew, membership of the Civil Defence Corps expanded to 330,000 by March 1956. Recruitment continued well into the 1960s, until the financial crisis of the mid 1960s resulted in the standing-down of the Civil Defence Corps in 1968. Nevertheless, civil defence planning remained active within the Home Office's Civil Defence and Common Services Department, whose responsibilities were passed to F6 Division of the Police Department in 1970, who in turn passed them on to the Fire and Emergency Planning Department in 1984, which became the Emergency Planning Division in July 1989.

2 Civil Defence Structures

2.1 Air Raid Precaution and Civil Defence Control Centres

In the Second World War, Divisional Control Centres and Sub-divisional Control Centres were often situated at first in the basements of town halls and civic buildings; later, purpose-built structures were often erected nearby or in parks. Second World War examples were generally rectangular, or sub-rectangular, single-storey brick structures with over-hanging flat concrete roofs, resembling large surface air raid shelters; Alverstoke, Gosport (Hampshire) is a Grade II-listed example (Fig 11). Others were semi-sunken or buried structures, for example, Queen's Gardens Kingston upon Hull, and Castle Grounds, York, and, from the Cold War era, Woodlands Park, Gravesend (Listed Grade II), built in in 1954. The design varied from site to site but the interior was usually entered via a blast baffle that led into an axial corridor with an emergency exit at the opposite end. Depending on the size of the centre, rooms off the corridor might include male and female dormitories and toilets, an air conditioning and generator plant room, district control room, sub-divisional control room, message or telephone and teleprinter room, liaison officer's room, controller's office and messenger's room.

Figure 11
Civil Defence Centre, Alverstoke, Gosport, Hampshire, completed February 1941 for the local control of the Civil Defence services. Listed Grade II.

2.2 Air raid shelters

These fall into four broad headings: Domestic, Communal, Public, and Industrial.

Domestic shelters (for household use)

include the Anderson shelter, the Morrison shelter, strutted rooms, trench shelters, surface shelters and nuclear shelters.

The Anderson shelter This was built of 14-gauge corrugated steel sheets, steel angle-iron frames, ties and channel irons. The 'standard' shelter was 6ft 6in (1.98m), by 4ft 6in (1.37m), by 6ft (1.83m) high and was intended to accommodate up to six persons. It was intended to be semi-sunken, the floor level sunk about 3ft with the earth spoil mounded over the structure to a minimum depth of 15in. Anderson shelters could be lengthened or shortened to accommodate between four and twelve persons. Not all were sunk into gardens; some were erected within the drawing rooms of larger houses, and some in the street, outside the front doors of terrace houses and revetted with stones or blocks of concrete. They also served in military contexts, as weapons stores, or even detachment shelters at searchlight sites. Post-war, many were issued as garden sheds in 'pre-fab' and council house estates.

The Morrison shelter This was a steel table measuring 6ft 4in long, by 3ft 10½in wide designed for internal use in family homes. Most were recovered post-war.

Strutted rooms These were usually basement adaptions, employing a corrugated sheeting or steel mesh lining supported by rolled-steel joists and angle-iron beams and struts. They were manufactured by companies like W.G. Allens & Sons (Tipton) Ltd. Most strutted room shelter steel was recovered as part of the post-war scrap drives, although examples do survive.

Trench shelters Trenches were typically 5ft wide and ideally at least 6ft deep, dug to a variety of plans. Various materials were used to support the trench sides and roof – corrugated sheeting, timber, brick, concrete panels, *in-situ* cast concrete, Xpamet steel reinforcement sheets, or even hessian sacking. Not all trench shelters were wholly below ground and some were semi-sunken and mounded.

Surface shelters These were generally built to accommodate between four and six people, sometimes singly, or in multiples of up to four, straddling property boundaries. Generally they were rectangular or square, with a concrete floor and 12in thick brick walls with a square emergency escape panel with bricks dry laid in weak cement in an end or side wall. The roof was usually a 6in thick reinforced concrete slab; however, occasionally some were built with a crow-stepped profiled brick roof. Small examples are frequently mistaken for coal houses or yard sheds.

Nuclear shelters No domestic nuclear shelter design was officially sanctioned during the Cold War, although Government advice leaflets were produced giving householders advice on the creation of temporary shelters. Any shelters that do exist were private commissions, architect designed or bought off-the-shelf, latterly from Swedish or Swiss civil protection companies. Accordingly, these structures are only likely to be found at affluent homes (Fig 12), and are relatively rare. They were designed to allow continuous

Figure 12
Taverham, Norfolk, entrance to a nuclear shelter built within a private garden.

Figure 13
Entrance to the communal air raid shelter at St Leonard's Court, Richmond upon Thames, Surrey, built in 1938 with a capacity to hold forty-eight persons. Listed Grade II.

occupation for at least two-weeks, to allow nuclear fall-out to decay to a safe level. Shelters were often sub-divided into separate rooms for cooking, washing, sleeping and living. Storage of food and water, the provision of some form of lighting and the filtration of air were essential, as was the means to dispose of human waste. Access was via a vertical shaft, or by a protected stairway; blast doors and blast baffles sometimes protected entrances. In a few exceptional cases house basements were designed as nuclear shelters, similar to many in former West Germany.

Communal shelters

Communal shelters were provided for the joint use of the occupants of a number of houses, flats, or streets where it was impracticable to provide shelter for the individual properties. The three most common types are trench, strutted basements, and surface shelters. In addition, some progressive private flat developments built during the 1930s were provided with a communal air raid shelter such as the Grade II-listed example at St Leonard's Court, Richmond on Thames (Fig 13).

Trench shelter Constructed on a much larger scale than domestic examples, these were usually dug in parks and open spaces, 4ft 9in to 7ft (2.14m) wide, approximately 30-58ft long and with a depth of not less than 6ft. Either at least 18in of earth was heaped over, or a 6in concrete roof was provided for overhead cover. Trenches were lined in various materials including parabolic concrete panel arches. Such shelters were designed to have a fifty-person capacity, were usually entered by steps at one end alongside an Elsan lavatory, and with an emergency escape hatch at the opposite end. Multiple units could be used to cater for larger numbers and this was achieved by a 'dog-leg' change of direction, linking two or more or 50-person capacity units. Trench shelters were supplied with rudimentary bunks or bench seating using brick or timber struts supporting timber planks or smooth asbestos cement sheets.

Surface shelters As the name suggests, these were built entirely above ground, and had a rectangular plan with a baffled porch entrance at one end and a dry-laid brick emergency panel at the other. The walls were usually built of reinforced brickwork not less than 13½in thick, or of 12in reinforced concrete, with a flat reinforced concrete roof of not less than 6in deep, vented by ventilation bricks. Many hundreds of thousands of 50-person surface air raid shelters were built in so-called 'single' or 'double' designs, and others were built with 25, 30, 100, or even 200-person capacities. Where the capacity was greater than 50-person, the interior was divided into 50-person units by staggered baffle walls. Surface shelters can still be found in isolation, or associated with factories, schools and hospitals, often serving as stores. Where fittings like seating, closet curtains, ventilation baffles, light fittings (electric lamps, wooden boxes for candles, or hooks for Tilley lamps) survive, this adds to their interest.

Figure 14
Steel strutted basement shelter beneath the Coronet Cinema, Notting Hill Gate. Listed Grade II.

Public shelters

Public shelters were adapted or constructed to provide shelter for people caught out in the street when an air raid warning was sounded. Initially, they took the form of trench shelters, railway arches, or strutted basement shelters in, for example, shops, church halls, and cinemas (Fig 14), and purpose-built surface shelters. Subsequently, the demand for shelter for large numbers of individuals staying overnight became ever more apparent during the night blitz. Cave and tunnel systems were adapted as shelters in places including Birkenhead (Wirral), Clifton (Bristol), Dover and Ramsgate (Kent), Nottingham (Nottinghamshire), Salford (Greater Manchester), etc. As noted above, as the blitz intensified, nightly 'trekking' led to the construction of surface shelters on the periphery of towns, provided with medical services and canteens.

After initial Government resistance to the use of London Underground stations as shelters, their nightly occupation resulted in a change of mind and they were opened as official deep shelters. Four disused tube stations – Aldwych, Leytonstone, Liverpool Street and Southwark – were re-opened specifically for this purpose. By late 1940, the value of the tube stations as shelters had been recognised and the government proposed to increase capacity by excavating ten purpose-built deep shelters alongside existing tube stations. The scheme was dogged by shortages and funding issues; eventually eight were completed, during 1942. Purpose-built deep shelters were also tunnelled on the periphery of London and at a number of towns where the geology lent itself to the task: at Birkenhead, the London Borough of Croydon, Epsom, and Epsom Downs (Surrey), Guildford (Surrey) and Portsmouth (Hampshire).

Industrial shelters

Industrial shelters were of all the types discussed above. Trench and Surface shelters often survive around the periphery of factory complexes as at Boots (Nottingham) and Empire Mills (Selby, North Yorkshire). Some shelters were built into the design of factory buildings from the outset, as at Park Works, Kingston upon Thames (listed Grade II), often functioning as corridors or washrooms until needed as shelters. These rooms were usually built to a higher specification than the remainder of the building, with thicker re-enforced brick or concrete walls and ceilings, baffle doorways, and high-set narrow Crittal windows glazed with Georgian glass to reduce injuries caused by flying glass. Other shelters were built on or under the surface of the factory floor, like a former car showroom in Chester which was converted into a shadow factory for the assembly of the front sections of Avro Lancaster bombers. Some shelters found a later use as store rooms, offices or workshops, as with the Grade II-listed examples at the Ecko works, Southend-on-Sea (Essex). Under-floor shelters ranged from those for 50 people to substantial complexes of inter-linked shelters, accommodating hundreds or even thousands, divided into sections, each served by warden posts and first aid rooms.

Where geology and situation permitted, tunnel shelters were also dug back into hillsides at sites such as the Henley Telegraph Works, Northfleet, and Shorts Brothers aircraft factory, Rochester (both Kent). Some pre-existing tunnels were also brought into service as underground factories, where protection was taken one stage further. For example, during 1940 concerns over the vulnerability of the Plessey factory in Ilford (London Borough of Redbridge) brought about the transfer of all of the manufacturing equipment and the staff to a five-mile length of 12ft (3.66m) diameter tube tunnels on the Central Line, with access for the 4,000 employees via Wanstead, Redbridge, and Gants Hill stations. Shortly after the completion of the move in March 1942, the empty Ilford factory did indeed suffer severe damage in a bombing raid.

2.3 Air Raid Warden Posts

Many ARP posts were in pre-existing buildings, the only evidence today being bricked-up windows, thickened walls, or faded signage. Most purpose-built warden posts (like the Grade II-listed example at Dover) resembled miniature surface air raid shelters, large enough for a bunk bed, a desk, and the storage of rescue equipment. In some cases, some were even attached to a surface air raid shelter (Fig 15).

2.4 Fire Watcher Posts

Fire watcher posts were provided to give fire watchers/wardens and/or bomb watchers ('Jim Crows') protection from the elements and from blast and bomb fragments, whilst permitting them to observe their surroundings. Fire watchers were responsible for reporting fire outbreaks and for dealing with incendiary bombs to prevent fire taking hold, whereas the Jim Crows monitored the approach of enemy bombers and if necessary sounded the alarm for workers to take shelter. Posts were usually sited on the roofs of factories or tall buildings and were relatively small brick or concrete structures, with flat concrete roofs and wide-splayed observation loops; most are secondary, with some being built onto or into pre-existing stair or lift heads. Factories built during the Second World War usually had integral fire watcher posts. Today these structures are frequently mistaken for defensive pillboxes (Fig 16).

Figure 15
Air raid warden post with brick blast wall protecting the entrance, with stirrup pumps to hand, and a diamond-set steel cover for the emergency escape panel in the end wall.

Figure 16
Firewatcher's Post added to the roof of H D Symons and Co Ltd, Park Works, Borough Road, Kingston upon Thames in 1941. Listed Grade II.

Figure 17
A rare surviving steel fire watcher's / air raid warden's shelter to the rare of the former Newcastle Breweries office 101-115 Percy Street, Newcastle upon Tyne, Tyne and Wear. Listed Grade II.

One- or two-man proprietary steel observation cupolas or shelters were produced by a number of companies including W.G. Allen and Sons of Tipton and Constructors Limited of Birmingham. The W.G. Allen observation cupola was a steel cylinder designed to be fitted on the roof of an ARP post and was accessed from below by a ladder, whereas the one-man Consol firewatcher's shelter had a similar appearance to a salt-cellar, tapering from a diameter of 4ft 3in at the base to 2ft 5in beneath a domed cap. The latter was erected in areas where there was no cover, and it was designed to resist damage from falling masonry and bomb fragments. It should be noted that the military also utilised these shelters as protected sentry posts. Surviving *in situ* examples of these steel shelters are extremely rare (Fig 17).

2.5 Gas Decontamination Centres and Cleansing Stations

The great fear engendered by the use of chlorine, mustard gas and phosgene during the First World War pervaded inter-war civil defence planning. There was no standard design of gas decontamination buildings, and like many other civil defence structures some re-used suitable existing buildings and can be found in places as diverse as council yards, parks or even a dairy. All purpose-built decontamination centres were built in reinforced brick or concrete with a flat over-hanging concrete roof, doorways at either end, and if windows are present, they were either high-set under the eaves of the roof or fitted with steel blast shutters. The structures housed an undressing area (either indoors or out), an eye douche, showers, treatment, dressing rooms and a boiler room with a combined chimney and header water tank tower. The walls and floors were often covered in white tiles to ease cleansing. If a centre was to serve both sexes, which was often the case, the structure would be divided by a central spine wall. Some centres had chutes in the end or side walls of the undressing rooms for the disposal of contaminated clothing into bins outside; these were often bricked-up if the building was given a new use.

2.6 Static Water Tanks or Emergency Water Supplies (EWS)

The provision of emergency water supplies to facilitate the rapid extinguishing of fires was an essential component of Civil Defence. A wide variety of designs and materials were used in their construction including rubber, corrugated steel, and pre-fabricated steel panel tanks (Braithwaite tanks). Most would leave little or no trace, being equally easy to assemble or dismantle. Most surviving tanks are at factories and remain in use. The basements of demolished bombed buildings were also pressed into use as static water tanks.

2.7 Air Raid Sirens

The electrically driven two-tone air raid sirens that were used during the Second World War were generally placed on public buildings. Where additional coverage was required in urban areas or in smaller rural communities sirens were placed on tall timber or steel posts, with a double-door cast-iron cable terminal pillar at the base. Such structures are increasingly rare. The sirens themselves were manufactured by a number of

Figure 18
A typical three-bay 'dry stuffs buffer depot at Bishop Lydeard station yard, Somerset.

Figure 19
Painted signage within the tunnel shelters, dug back into the chalk cliff at Henley Telegraph Works, Gravesend, Kent.

companies including Gent of Leicester, Carter of Nelson, Castle Castings and the Klaxon Signal Company. The sirens continued in operation throughout the Cold War, only being decommissioned in 1993. However, some remain 'live' at large industrial complexes like oil refineries, and others have been retained for flood warning.

2.8 Buffer Depots

During the Second World War the supply of food, fuel and raw materials came perilously close to breaking down on a number of occasions (Fig 18). This persuaded the Government to establish storage depots from late 1941. These were sited with great care to ensure good road access and, wherever possible, to adjoin either a railway line and/or a canal. With the advent of nuclear weapons the depots gained even greater strategic importance, as it was essential that all regions had sufficient stocks of food, fuel and raw materials should other parts of the country be devastated by atomic blasts and nuclear fallout.

The depots included 47 cold stores, 16 grain silos, grain stores located on disused military airfields, 160 general purpose depots and some 41 semi-underground fuel depots. The general purpose store is the most readily recognisable and numerous among the standard designs: 214ft x 120ft rectangular, triple-pile single-storey sheds with gabled asbestos cement-clad roofs. The number of sheds varied from depot to depot, sometimes occurring with curved profile Romney huts as at Quainton Station (Buckinghamshire). Some general purpose depots like Whitley Bridge (North Yorkshire) were co-located with a grain silo.

2.9 Signage

Painted and metal signage was commonplace during the world wars and Cold War but surviving examples are now very rare. The commonest painted signs externally directed the viewer to air raid shelters or emergency facilities (Fig 19). Internally, such signage tends to survive, if at all, in low-use areas such as basements or stores, or in former air raid shelters. Such signage can add markedly to the historic interest of a civil defence structure.

2.10 Blackout provision

Second World War blackout measures included small porches with plain side walls at factories and offices which functioned as light baffles, preventing a flash of light when an external door was opened at night. Window shutters took a number of forms that included timber frames into which a plywood panel could be fixed, hinged internal plywood or timber shutters, and external steel shutters that doubled as blast and bomb protection.

2.11 Graffiti

Wartime graffiti are commonplace in air raid shelters and can have historic interest. Semi-official artwork was sometimes encouraged in air raid shelters, particularly those at schools and factories.

2.12 Miscellaneous Structures

The sheer range of building designs, types and construction prevents the consideration of all Civil Defence structures in this document. Those not here discussed (and this not an exhaustive list) include include National Fire Service garages and stations, ambulance stations, mortuaries, rescue depots, Civil Defence mobile column depots, dressing stations, first-aid ambulance depots (St John's Ambulance and Red Cross), gas mask stores, rest centres, national camps, civil defence headquarters, rescue training sets, civil defence colleges and railway control centres.

2.13 Cold War Central Government Emergency headquarters

Great Britain was divided into twelve Civil Defence regions each administered by a Regional Commissioner, ten of which were in England. The commissioners were responsible for co-ordinating all local civil defence activities and were usually accommodated in the strengthened basements of large houses that contained operations, map, message, teleprinter and typing rooms.

Following the re-establishment of the Civil Defence Corps, the Commissioners' role increased to include the assumption of full state power after a nuclear attack. This required a large permanent staff from different government departments. Work commenced on building thirteen war rooms in England, four of which were in Region 5 – London. The war rooms (Fig 20) were designed to resist a near atomic explosion, and could operate for at least two weeks using a diesel generator to provide power, with an air conditioning plant room for filtered and cooled air. The rooms were rectangular with 4ft 10in thick reinforced concrete walls, and a 5ft thick roof with protected concrete vents at one end. Internally, the war rooms were similar to the wartime regional headquarters, but were on much a larger scale on two floors and included offices for different government departments, scientific advisers, the emergency services, a military liaison officer, canteen and welfare rooms, and dormitories and facilities for different sexes.

Figure 20
Bristol War Room, map room showing curved Perspex viewing windows.

In 1953, the Soviets detonated a hydrogen bomb, with yields in the megaton range that could destroy great swathes of land and could affect wider areas for greater lengths of time. As a consequence, the post-explosion recovery period would be greater and each region would have to operate autonomously for longer, putting a great strain on the relatively small staff working in a war room. To provide for the greater manpower necessary to support the Commissioners, a system of Regional Seats of Government (RSG) was established during the early 1960s, with the London war rooms abandoned altogether. The increase in staff size (up to 400), including those from the BBC, resulted in new purpose-built war rooms being built adjoining the former war rooms at Cambridge and Nottingham, while other RSGs were established in abandoned RAF ROTOR bunkers as at Bolt Head (Devon), in tunnels beneath Dover Castle (Dumpy, in Kent), and in a former underground aircraft factory at Drakelow (Worcestershire).

No sooner had the RSGs come into service, it was realised that conditions after a nuclear strike might be even more fractured than first thought and a decision was taken to establish 25 Sub-Regional Headquarters (SRHQ), each accommodating a staff of up to 200 to assist during the recovery phase. By the end of the 1960s a number of SRHQs had been established in former army anti-aircraft operation room bunkers such as Frodsham (Cheshire). Others were constructed in purpose-built basements beneath new government buildings, as at the Civil Service Commission HQ, Basingstoke (Hampshire). This organisational structure was changed once again during the early 1980s, when England was divided into nine defence regions. Here the RSGs became Regional Government Headquarters (RGHQ) and the SRHQs were only to be occupied if the RGHQ was non-functional. The RGHQ oversaw the running of the region, with County Main Controls and back-up County Standby Controls, administering District Controls. Not all of the RSGs were re-occupied and a number of the RGHQs were established by altering and re-building former RAF bunkers as at Hack Green (Cheshire), Shipton-by-Benningborough (North Yorkshire) and Goosnargh (Lancashire); one was even built within a former cold store buffer depot at Hexham (Northumberland).

2.14 County, Borough and District Council headquarters

Below the central government headquarters the organisation of civil defence mirrored local government structures. From the late 1930s all local authorities down to district level were required to undertake civil defence planning. Dedicated civil defence headquarters were usually established in the basements of existing council buildings. Typically, these areas would be strengthened and protected from the ingress of poison gas. Council buildings constructed from the late 1930s often incorporate purpose-built civil defence headquarters. During the Cold War, earlier headquarters were often re-used and modified to counter the effects of nuclear weapons. From the 1960s, as the services abandoned bunkers these were often modified by local authorities, such as the former Anti-Aircraft Operations Room at Frodsham, Cheshire; (listed Grade II; Fig 21). During the 1980s there was renewed interest in civil defence and a number of local authorities constructed purpose-built emergency centres.

Figure 21
Frodsham Cheshire, the gallery of the operations room of the Cheshire County Standby Emergency Centre.

3 Further Reading

There is only a limited literature on civil defence during the First World War. One exception is Neil Hanson, *First Blitz – The Secret German Plan to Raze London to the Ground in 1918* (2009). Edwin A. Pratt's *British Railways and the Great War* (1921) is one of the best accounts of air raid procedures, including shelters.

Good general introductions to civil defence in the Second World War are Mike Brown, *Put That Light Out! - Britain's Civil Defence Services at War 1939-1945* (1999) and Ian Brown and others, *20th-Century Defences in Britain: An Introductory Guide* (1995). Many books have been written on the home front in the same period, and while the majority focus on events, people and uniforms rather than structures they do nevertheless put the buildings into context. They include Juliet Gardiner, *Wartime Britain 1939-1945* (2005); E.R. Chamberlin, *Life in Wartime Britain* (1972); Norman Longmate, *How We Lived Then – A History of Everyday Life During the Second World War* (2002); Charles Whiting, *Britain Under Fire* (1999); Martin J. Brayley *The British Home Front 1939-45* (2005); and Stephen Wade, *Air Raid Shelters of the Second World War: Family Stories of Survival in the Blitz* (2011). Useful discussions of civil defence structures appears in the three-volume *The Blitz – Then and Now*, edited by Winston Ramsey (1987-90). These include 'Civil Defence' (vol1, pages 99 – 107), 'Life in the Shelters' (vol 2, pages 112-21), and 'The Formation of the National Fire Service' (vol 3, pages 58 – 65). The most scientifically informed and contextually interesting pre-war analysis of Civil Defence preparations is given in J.B.S. Haldane, *A.R.P.* (1938). Many official manuals, handbooks and pamphlets were published throughout the period under consideration, often available at on-line auction sites, and some republished as facsimiles. English Heritage's Monuments Protection Programme (MPP) commissioned eleven reports entitled *Twentieth-Century Fortification in England* in 1994. Colin Dobinson, *Twentieth-Century Fortification in England*, Volume VIII - 'Civil Defence in World War II – Protecting England's Civil Population 1935-45' (1999), provides a particularly good insight into the content of, and the understanding to be gained from, the many Civil Defence documents held at the National Archive, Kew. English Heritage also commissioned a study of railway control centres: P. Francis, P. Bellamy and G. Crisp 20th-Century Air-raid Precaution Railway Control Centres' (Airfield Research Group, 2014). These last two reports are available through Historic England's Archive https://archive.HistoricEngland.org.uk/ and most Historic Environment Records.

The best introduction to the historical background to Cold War civil defence is Matthew Grant, *After the Bomb: Civil Defence and Nuclear War in Britain 1945-68* (2010). Its physical remains are described by Paul Ozorak, *Underground Structures of the Cold War* (2012), Bob Clarke, *Four Minute Warning: Britain's Cold War* (2005), Wayne D. Cocroft and Roger J. C. Thomas, *Cold War, Building for Nuclear Confrontation 1946-1989* (2003), Nick Catford, *Cold War Bunkers* (2010), and N. J. McCamley, *Cold War Secret Nuclear Bunkers: The Passive Defence of the Western World During the Cold War* (2013).

3.1 Contact Historic England

East Midlands
2nd Floor, Windsor House
Cliftonville
Northampton NN1 5BE
Tel: 01604 735460
Email: eastmidlands@HistoricEngland.org.uk

East of England
Brooklands
24 Brooklands Avenue
Cambridge CB2 8BU
Tel: 01223 582749
Email: eastofengland@HistoricEngland.org.uk

Fort Cumberland
Fort Cumberland Road
Eastney
Portsmouth PO4 9LD
Tel: 023 9285 6704
Email: fort.cumberland@HistoricEngland.org.uk

London
1 Waterhouse Square
138-142 Holborn
London EC1N 2ST
Tel: 020 7973 3700
Email: london@HistoricEngland.org.uk

North East
Bessie Surtees House
41-44 Sandhill
Newcastle Upon Tyne
NE1 3JF
Tel: 0191 269 1255
Email: northeast@HistoricEngland.org.uk

North West
3rd Floor, Canada House
3 Chepstow Street
Manchester M1 5FW
Tel: 0161 242 1416
Email: northwest@HistoricEngland.org.uk

South East
Eastgate Court
195-205 High Street
Guildford GU1 3EH
Tel: 01483 252020
Email: southeast@HistoricEngland.org.uk

South West
29 Queen Square
Bristol BS1 4ND
Tel: 0117 975 1308
Email: southwest@HistoricEngland.org.uk

Swindon
The Engine House
Fire Fly Avenue
Swindon SN2 2EH
Tel: 01793 445050
Email: swindon@HistoricEngland.org.uk

West Midlands
The Axis
10 Holliday Street
Birmingham B1 1TG
Tel: 0121 625 6870
Email: westmidlands@HistoricEngland.org.uk

Yorkshire
37 Tanner Row
York YO1 6WP
Tel: 01904 601948
Email: yorkshire@HistoricEngland.org.uk

This page is left blank intentionally

4 Acknowledgements

Images

© Historic England

Figures 6, 9-11,

© Other

Front cover: The National Archives, TNA HO205/156

Figure 1: Picturepoint-TopFoto

Figure 2: TNA HO 205/149

Figures 3, 5, 7, 8, 18, 20-21: Roger J. C. Thomas

Figure 4: Crown copyright HMSO

Figure 12: Trish Roberts

Figure 14: Mark Butler

Figure 15: Crown copyright, Ministry of Information

Figures 13, 16: Patience Trevor

Figure 17: Neil P Elborn

Figure 19: Veronica Fiorato

Every effort has been made to trace the copyright holders and we apologise in advance for any unintentional omissions, which we would be pleased to correct in any subsequent editions.

Historic England

We are the public body that looks after England's historic environment. We champion historic places, helping people understand, value and care for them.

Please contact
guidance@HistoricEngland.org.uk
with any questions about this document.

HistoricEngland.org.uk

If you would like this document in a different format, please contact our customer services department on:

Tel: 0370 333 0607
Fax: 01793 414926
Textphone: 0800 015 0174
Email: customers@HistoricEngland.org.uk

All information and weblinks accurate at the time of publication.

HEAG145
Publication date: October 2016 © Historic England
Design: Historic England

Printed in Great Britain
by Amazon

64240627R00015